SHERALIEVA RISOLAT ISLOM KIZI
TASHPULATOV SALIKH SHUKUROVICH

Creation of special technology of fabric production

SHERALIEVA RISOLAT ISLOM KIZI
TASHPULATOV SALIKH SHUKUROVICH

Creation of special technology of fabric production

With properties of electrical conductivity on the basis of jacquard functional knitting

ScienciaScripts

Imprint
Any brand names and product names mentioned in this book are subject to trademark, brand or patent protection and are trademarks or registered trademarks of their respective holders. The use of brand names, product names, common names, trade names, product descriptions etc. even without a particular marking in this work is in no way to be construed to mean that such names may be regarded as unrestricted in respect of trademark and brand protection legislation and could thus be used by anyone.

Cover image: www.ingimage.com

This book is a translation from the original published under ISBN 978-620-6-18288-7.

Publisher:
Sciencia Scripts
is a trademark of
Dodo Books Indian Ocean Ltd. and OmniScriptum S.R.L publishing group

120 High Road, East Finchley, London, N2 9ED, United Kingdom
Str. Armeneasca 28/1, office 1, Chisinau MD-2012, Republic of Moldova, Europe
Printed at: see last page
ISBN: 978-620-6-98937-0

Contents

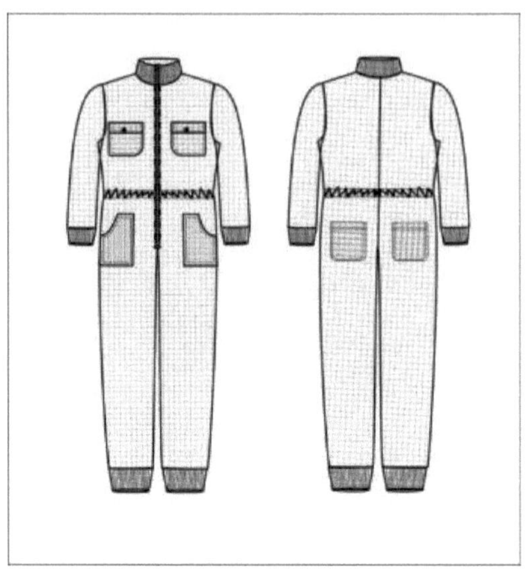

INTRODUCTION

According to the decree of the President of the Republic of Uzbekistan 7 July 2017 No PF-4947 "Development Strategy of the Republic of Uzbekistan 2017-2021" and No PF- 5285 dated 14 December 2017 "Advanced development of embroidery and embroidery production" Development Strategy of the Republic of Uzbekistan Uzbekistan Uzbekistan in 2021. The research of this dissertation will contribute to the implementation of the decision of September PQ-4453 "Programme of measures for the development of light industry and the production of finished goods" and other legal and regulatory documents related to this activity to a certain degree.

The textile and garment and knitwear industry of Uzbekistan is one of the leading and dynamically developing industries. It plays an essential role in solving state tasks and meets the vital interests of many regions. It contributes to the harmonious development of the regions, provides employment and improves the well-being of the population, and assists in the establishment and development of small and private businesses.

The most important area of our internal reserves and capabilities was a step-by-step increase in the depth of processing of domestic raw materials, as well as the expansion of the volume and range of high value-added products.

Over the past 27 years, the industry has become one of the leaders both in attracting foreign investment and in exporting high value-added products. Today, the industry is represented by a wide export range - from yarn to finished garments and knitwear.

The light industry of Uzbekistan is one of the leading and dynamically developing industries. The annual growth of the industry's output in recent years has been approximately 18 per cent and exports 10 per cent.

The largest increase in value added in the cotton textile value chain is in the area of ready-made knitwear and garments.

In light of the fulfilment of the set tasks, joint stock companies and enterprises of the Association in 2018 ensured the production of consumer goods for 7.5 trillion UZS, the growth rate to the corresponding period of the last year was 128.6%, the volume of industrial production amounted to 13.2 trillion UZS with a growth rate of 116.5%.

The main focus was on increasing the production of finished goods with high added value, the share of which exceeded 57% of total industrial production (against 52% in 2017).

CHAPTER 1

DEGREE OF STUDY OF THE PROBLEM OF DESIGNING MODERN OVERALLS FOR ELECTRICIANS

Foreign scientists Makoto Akashi (Japan), scientist of the University of California William W. Nazaroff (USA), Castillo J., Cubillos A. (Colombia) among the tasks of studying the properties of materials to confirm and document the properties of electrical conductivity and protective clothing against electric field, Azhar Ali, Patrick (India), E.V.Romanov, E.Y.Surzhenko, P.P.Koketkin, E.B.Koblyakova, I.V.Cherunova, 3.S.Chubarova, I.N.Savelyeva (Russia) and others on the scientific and scientific basis of the production of protective clothing on the launch .

Б. I. Sazhin, G. M. Bartenev, G. A. Lushchekina, P. L. Gefter, E. V. Silchenko, A. A. Aslanian, D. N. Akbarov, S. Sh. Toshpolatov, B. Kh.

As a result of this work formed a system-forming methodological basis for the design of special clothing for various purposes, factors and criteria of compatibility with the real conditions of operation "man - clothing - atrophy - environment", nomenclature, design ergonomic league. production, etc. or quality of fabric and its materials for the production of a number of quality products and techniques, but in hot and cold conditions of our republic to use in the production of workwear that reliably protects workers from the effects of energy supply, but in the hot and cold conditions of our republic to use in the production of special clothing, reliably protects workers from the effects of energy supply. No research has been carried out on the development of alternative textile materials and production of special clothing that are an alternative to ketil and have high electrical conductivity and protection from electric fields.

CHAPTER 2

THE SAFETY OF EMPLOYEES IN ANY FIELD IS PARAMOUNT

In the world, the use of energy-saving technologies and equipment for the production of garments occupies one of the leading places. Globally, research and development activities are carried out to create knowledge-intensive equipment and technologies, to identify important factors affecting technological processes. Creation of new types of high-quality and competitive clothing, including the creation of special clothing that meets hygienic, ergonomic, operational and aesthetic requirements, the creation of rational design while ensuring its reliability, the development of optimal technological parameters, increasing the types of special clothing and improving quality indicators is an actual scientific activity is one of the practical issues.

In our republic, comprehensive measures are being implemented to improve the quality and competitiveness of light industry products, integrated processing of raw materials into finished products, and certain results are being achieved. The Development Action Strategy of the Republic of Uzbekistan for 2017-2021 defines important tasks, including "...increasing the competitiveness of the national economy, ...reducing the consumption of energy and resources in the economy by widely introducing energy-saving technologies in production..."[3]. In the implementation of these tasks, including using progressive methods of ergonomic design based on the principles of complex analysis of the system "man-clothing-external environment", creating reliable and high-quality protective clothing for workers in various spheres of activity, including in the energy sector, including electrical conductivity and protection from electric field, the development of special clothing from special fabrics is becoming increasingly important.

CHAPTER 3

SPECIAL CLOTHING AS A GUARANTEE OF HEALTH OF AN ELECTRICIAN'S EMPLOYEE

Creation of new types of high-quality and competitive clothing, including the creation of special clothing that meets hygienic, ergonomic, operational and aesthetic requirements, the creation of rational design while ensuring its reliability, the development of optimal technological parameters, increasing the types of special clothing and improving quality indicators is an actual scientific activity - is one of the practical issues [2,1]

Today in Uzbekistan, employees of electricity supply companies constitute a large professional group. Joint Stock Company "Regional Electric Networks" was established in accordance with the Decree of the President of the Republic of Uzbekistan dated 27 March 2019 "On the strategy of further development and reforming of the electric power industry of the Republic of Uzbekistan". management.

JSC "Territorial Electric Networks" has 209 power supplying enterprises, electricity is supplied to consumers of the Republic of Uzbekistan through 1623 substations, 79,122 transformers, 252.6 thousand kilometres of power transmission lines [8, 26.] .

A classification analysis of the occupations of district electricians has identified the following activities [9, 206] :

1. Electricians who maintain power plants;
2. Electricians repairing electrical wires;
3. Electricians who maintain industrial equipment;
4. Electric welders.

Workers in these occupations may encounter the following electrical injuries while working.

Electrical trauma is the effect of electric current on the body with damage to tissues, skin and muscles. Electrical injuries in the form of burns are particularly dangerous. Electrical burns occur when the human body comes into contact with the live part of an electrical device or electric arc [10].

GOST 12.1.002-84. In accordance with SSBT [18.1-7 bb] the term "Electrical safety" means a system and means of organisational and technical measures to protect people from harmful and dangerous effects of electric current, electric arc, electromagnetic field and static electricity.

Hot and dry weather conditions exacerbate the existing problem of electrostatic hazards of low conductivity synthetic materials. According to statistics, 27 % of electrostatic explosions were caused by static electricity, 40 % of which resulted in death [11, 3p].

The danger to life and health of workers in an accident involving travelling electric discharge is as follows:

- this accident happens suddenly, it's almost impossible to predict;
- The accident causes a fire that burns the worker's clothing and causes serious damage to the human body by burning the skin as a result of the intense heat flow;
- When exposed to high temperatures, synthetic parts of work equipment and clothing melt and stick to human skin, additionally causing serious injuries;
- the occurrence of an electrical discharge can be fatal due to the release of a large concentration of energy in a short time and human exposure.

The basis of an electrician's outfit is the work suit. The main thing is that it should be comfortable to work in it.

The colour of overalls for an electrician is not fundamental: you can pick up for work blue, green, dark grey and any other set.

What elements of workwear will be comfortable for working with electrical equipment:

Jacket made of breathable material with long sleeves. The design had enough pockets for tools. For safety reasons, it is better to choose a jacket with a minimum of metal fittings - for example, with textile Velcro fasteners instead of buttons.

Trousers or overalls with a comfortable cut. Deep pockets are also important here, as well as an elasticated waistband and height adjustment (if you choose a dungaree).

Summer waistcoat that can replace a jacket when working in very warm environments. The waistcoat protects against dirt, allows you to place the necessary tools in the compartments, and at the same time does not constrain your movements.

CHAPTER 4

DANGER OF ELECTROMAGNETIC RADIATION BY WAVES ARISING IN THE ELECTROMAGNETIC MEDIUM

Thus, an important negative factor in the field of electric power industry is the electromagnetic radiation hazard arising in the working environment, protection against which determines the basic requirements to the characteristics and quality of appropriate workwear.

Depending on the sign, magnitude and location of the electrostatic charge, the electric field created by it can have a negative impact on the human body and even threaten its health [12,26.; 13, pp. 184-190].

Accumulated electric charge can cause injuries and damage due to leakage of electric current through the body, sparking or contact with harmful substances in the environment. Prolonged exposure of the human body to an external electrostatic field can cause functional disorders of the nervous and cardiovascular systems. The totality of physiological effects of the electric field on a person and the consequences arising from them are conditionally divided into four types [15, 16, 2-3 p.].

Estimated distribution of electrical injuries in industry by the specified types of injuries: 20% - local electrical injuries; 25% - electric shock; 55% - mixed injuries, i.e. simultaneous local electrical injuries and electric shock [17, 26] .

Local electrical trauma is a vivid manifestation of damage to body tissues, including bone tissue, caused by the action of electric current or electric arc. Most often these are superficial injuries, i.e. damage to the skin, sometimes other soft tissues, as well as ligaments and bones. As shown, about 75% of electrocution cases are localised electrical injuries. According to the types of electrical injuries, these cases are distributed as follows [18,2 p.] :

1.2- table

Cases of electric shock of people by type of electric trauma

№	Types of electrical faults.	%
1	Electrical burns	40
2	Electrical (contact) burns	7
3	Metallisation of leather	3
4	Mechanical damage	0,5

5	Electroophthalmia	1,5
6	Electric shock	23
	Total :	75

Electric burns are the most common electrical injury: burns occur in the majority (63%) of victims of electric shock, one third of them (23%) are accompanied by other injuries - marks, skin metallisation, ophthalmia [18,2 p.] .

Electrical burns occur when the voltage in electrical appliances does not exceed 2 kV. At high voltages, an electric arc or spark usually occurs, resulting in another type of burn, scalding [18,36.] .

Skin metallisation - tiny particles of metal melted under the action of an electric arc penetrate the upper layers of the skin, and the victim feels a burning pain due to the heat of the metal entering the skin. Metallisation of the skin is observed in 10% of victims of electric shock [18,3 p.] .

Mechanical injury occurs as a result of a sudden forced contraction of muscles under the influence of a current passing through the human body, which can result in ligaments, skin, vessels and nerve tissue being cut, joints being dislocated and bones being broken. Mechanical injuries are very rare (1.0% of electrocution victims) [18,3 p.] .

Electro-ophthalmia (from Greek ophthalmos - eye) - inflammation of the outer membranes of the eye and conjunctiva (mucous membrane covering the eye), caused by the influence of a strong stream of ultraviolet rays, which are absorbed by body cells and cause chemical changes in them. Electro-ophthalmia occurs in about 3% of victims.

Electroshock is a severe neuro-reflex reaction with a specific response to electric current, which leads to profound disorders of blood circulation, respiration, metabolism and others. In shock, the nervous system stops and weakens, blood pressure drops sharply, pulse slows or increases, respiration shortens or increases, there is an oppressed state The shock state can last from tens of minutes to several days. After that, a person may die as a result of complete disappearance of vital functions or recover as a result of timely active medical intervention [18,4 p.] .

The information of classification of occupations in the Republic of Uzbekistan by electric power supply professions is presented in Table 1.3 [19, 46.]. As a result of analysing the topography of overalls issued to the electric power industry workers, the following influencing factors were determined:

- electricity: exposure to electricity and sparks;
- Mechanical: moving parts of production equipment, sharp edges and uneven surfaces;
- physical-mechanical: friction and contaminants cause damage to the protective elements (metallised fibres) of the material.

1.3rd Adv.

Classification of electrical supply occupations and types of hazardous exposures

groups	Name of class	Types of hazardous impacts
1	Electricians repairing electrical wiring	Electrical sparks, electric shock, electric field, mechanical effects, stereotypical work behaviour.
2	Electricians servicing industrial equipment	Electrical sparks , dustiness of air at the workplace, additional thermal effects from increased temperature of equipment surfaces, vibration and noise at the workplace, stereotyped actions during the shift
3	Electric arc welders	Electrical sparks , electric arc, dustiness of the air at the workplace , increased level of vibration and noise at the workplace, stereotyped actions during the

		shift

The uniforms of these employees are distinguished by the voltage value at the workplace. Protective suits are designed for the voltage of electrical equipment. To date, the following protective suits are issued for the employees of RECs of Uzbekistan (Fig. 1.1).

Figure 1.1. Overalls of employees of Uzbek JSC "Territorial Power Grids" are available

Thus, based on the above, providing power supply workers with special electrically conductive clothing is an urgent task and requires a comprehensive approach to its design. Therefore, at the next stage of our work we analysed the protective clothing issued to power engineers today and the external factors affecting them.

CHAPTER 5

ANALYSING RESEARCH AND DEVELOPMENT WORK ON DEVELOPMENT AND PRODUCTION

SPECIAL ELECTRICALLY CONDUCTIVE CLOTHING

The works of scientists from a number of foreign countries and CIS countries are devoted to the development and production of special electrically conductive clothing. A group of Japanese scientists together with Abel J. Gonzales, Makoto Akashi, John D. Boyce Jr. in their work they put forward the problem of protection of workers from radiation exposure, developed a programme of scientifically based requirements for the origin of radiation hazard, measurement of radiation exposure, assessment of the value of internal exposure, means of individual protection from radiation, as well as special clothing [20, 497-571 p.].

Authors Castillo J., Cubillos A. Workers of the Colombian electric power industry have developed ergonomic principles of designing special clothing [21, 623-627 p.].

Works of Korean scientists Lin LX, Wang YY, Lee J. are devoted to the issue of functional design and ergonomic evaluation of special clothing of rescuers for emergency situations, including search and rescue work in earthquakes [22, 348-355 p.].

Indian scientists Azhar Ali, Patrick recommend using nanocomposites in the materials of special clothing protecting against radiation. In their opinion, a high level of insulation can be achieved due to the large interface surface area between nanocomposites and polymer, and this approach is superior to previous ones [23, pp. 2814-2819].].

Professor V.E. Romanov was the first to put forward a systematic approach to the design of special clothing, developed its theoretical and methodological foundations for further improvement [24, 54 p.].

E.A. Surzhenko proposed a new approach to the ergonomic design of special clothing, based on the biokinematic analysis of the interaction between the elements of the system "man-clothing" [25, 35-36 p.]. [25, 35-36 c.].

E.V.Silchenko has developed a new antistatic fabric protecting a person from electromagnetic radiation, significantly attenuating electromagnetic and electric fields, as well as meeting the requirements for surface density, thickness and parameters of the fabric structure[26, 112 p.].

I.V.Cherunova has developed an improved design of special electrically conductive clothing for oil workers. In this work, the characteristics influencing the characteristics of a package of overalls for cold climatic conditions are studied by methods of mathematical analysis; The heat-resistant properties of clothing packages were studied using special intermediate materials [27, 143-145 p.].

A. V. Merkulova proposed a new method of developing special electrically conductive clothing protecting from low temperatures, and for the purpose of designing electrically conductive clothing was developed

regional climatic map of the northern regions of Russia. A mathematical model of electrification of a set of workwear materials has been developed, allowing to determine in advance its antistatic protective properties at the design stage. For protection against low temperatures the methodology of designing special electrically conductive clothing has been improved, a special system of warning about the strengthening of electrostatic field inside overalls has been developed [28, 113-123 p.].

I.V. Kurenova proposed criteria for assessing the thermal field to improve the thermal safety of clothing; the dependence of thermal conductivity of synthetic intermediate insulating layer of clothing on the mass fraction of oil is determined; the method of acoustic-emission analysis of oil absorption dynamics is proposed; the mathematical model of temperature field distribution in the layers of the system "man-clothing-environment" is developed [29 p. 14-17]. [29, c. 14-17].

И. N. Ivashchenko determined the real working conditions during oil production in the oilfields of the southern region of Russia, developed a technology for designing ergonomic design of special clothing taking into account the dynamics of workers' movements, thermophysical parameters and conditions of use. . In addition, the design of protective clothing with adjustable thermal insulation in the range of variable energy inputs and air temperature was theoretically justified and experimentally confirmed [30, 98-114 bb].

Y.O. Shultz considered the issues of developing the theoretical basis for designing special heat protective clothing based on the calculation of the unstable temperature field from the source of impact on the human body. In this work a new method of obtaining special material with polymer-metallised coating was developed, as well as devices and methods for determining the thermal resistance and protective properties of materials [31, 3-8 p.].

Development of antistatic fabrics is also devoted to scientific works of a number of Uzbek scientists[32-34], including D.N. Akbarov's work "Giving antistatic properties to textile materials" [32, 97-99 p.] considers the issue of providing antistatic properties of special purpose fabrics.

R.D. In Akbarov's research paper "Development and study of properties of special fabric with specified electrophysical characteristics", the physical and mechanical properties of electrically conductive metallic fibres are based on almost the same properties as the original nitron fibre. The author determined that the most optimal is a mixture of 60% cotton and 40% electrically conductive fibres, and it was proved that the addition of electrically conductive fibres to cotton fibres improves their physicomechanical and operational properties [33, P. 3-19].

B.H. Baimuratov in the scientific work "Creation of multifunctional electrically conductive textiles

Baimuratov B.H. In the scientific work "Creation of multifunctional electrically conductive textile materials and their properties" on the basis of dependence of distance between electrically conductive threads in the fabric on shielding efficiency it is proved that this distance a=4-6 mm is rational. and provides 50-fold reduction of shielding efficiency. The fabric created by the author is recommended for making special covers and clothes protecting workers from electromagnetic radiation (radar devices, microwave equipment) and high-voltage electric fields (power lines, transformers, computer equipment, etc.) [34, 26 -28 bb].

Thus, as a result of the analysis it was found that the majority of scientific works are aimed at the development of fabrics with antistatic properties and improvement of their physical-mechanical and operational properties, the number of scientific works aimed at the production of electrically conductive overalls is insufficient, and all of them are carried out for cold climatic conditions. During the study of scientific developments in the field of workwear design it was found that scientific research on the design of electrically conductive workwear for district power engineers taking into account the climatic conditions of the Republic of Uzbekistan has not been carried out. Taking into account modern labour conditions of workers of regional electricity supply of Uzbekistan, hazardous and harmful production factors negatively affecting the health of workers, and the design and development of quality special clothing from materials with protective properties, the main was determined the direction of the thesis.

CHAPTER 6

BASIC PHYSICAL, MECHANICAL AND CONDUCTIVE PROPERTIES AND APPLICATION OF TEXTILE MATERIALS FOR ELECTRICALLY CONDUCTIVE WORKWEAR

To date, the domestic market of our country lacks fabrics and special clothing adapted to the working conditions of workers in various spheres with a level of protection and safety . Metal fibre fabrics (Fig. 1.2) and products made of them are now widely used all over the world, including antistatic clothing, heat-reflective suits, filter screens, clothing for firefighters and rescuers, insulating coatings for special rooms to prevent the spread of information, etc.

Antistatic fabrics are used in the fields of power engineering, aviation, astronautics, shipbuilding and medicine. The material properties allow them to be used as protective equipment in high voltage environments.

B.N. According to Gusev, he proved that the electrical efficiency of a cotton cloth sample from friction is three times lower than that of a capron cloth sample under conditions of relative humidity PH 50% and temperature +21 °C . . Similar electrical charge on static charge of the polyester fabric sample showed a significant effect on reducing the electrical conductivity . This is due to the fact that cotton fibre has high hygroscopic properties as the tendency of cotton fabrics to become static depends on the air humidity. At low air temperature and at zero humidity, cotton fabrics lose moisture and can accumulate the same electrostatic charge as fabrics made of synthetic fibres [35, 42-47 p.].

Measurement of electrical resistance of electrically conductive overalls in different climatic conditions GOST 10688-75 [36, 3-6 bb.]is carried out in accordance with the Measurement of electrical properties of materials used in hazardous areas with air humidity of 50 % and above is not applied abroad today, as it may jeopardise safety. The most optimal conditions for testing are a relative humidity of 25 % and a temperature of 23°C.

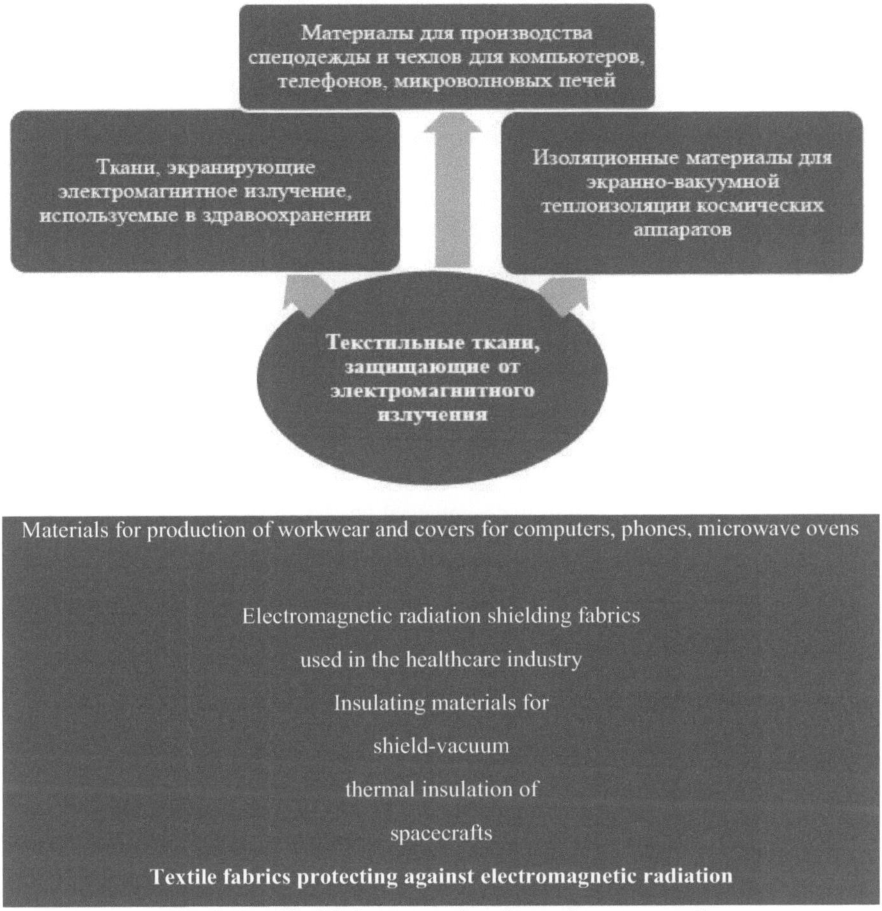

Материалы для производства спецодежды и чехлов для компьютеров, телефонов, микроволновых печей

Ткани, экранирующие электромагнитное излучение, используемые в здравоохранении

Изоляционные материалы для экранно-вакуумной теплоизоляции космических аппаратов

Текстильные ткани, защищающие от электромагнитного излучения

Materials for production of workwear and covers for computers, phones, microwave ovens

Electromagnetic radiation shielding fabrics used in the healthcare industry

Insulating materials for shield-vacuum thermal insulation of spacecrafts

Textile fabrics protecting against electromagnetic radiation

Fig.1. Textile fabrics protecting from electromagnetic radiation Tex-Centre Ltd.

Radio communication, electricity, radar, electrical engineering, science-intensive technologies - this is an incomplete list of areas where people need to work under the influence of this energy. Currently, the most convenient way to protect human devices from energy impact is fabrics made of metal fibre filaments.

Tex-Centre and NPP Tehnostil produce textile fabrics that protect against electromagnetic radiation. LLC "Tex-Centre" produces textile fabrics that protect against electromagnetic radiation (Fig. 1.3): insulation materials for spacecraft, radiation-protective clothing, cases for computers, telephones, microwave ovens [37,

1-2 p.].

In the production of metal fibre products "Metacron" of "NPP Tehnostil" company is used double-sided packaging. As a coating are used materials up to 12 microns thick, containing turlitol composites (polyester, polyamide, glass, basalt, aramid, etc.) [38, 1 p.] .

Designed for covering computers and microwave ovens, it has a light shielding capacity of up to 35 dB and the fibre composition is 90% polyester, 9% copper, 1% silver. Silver Elastic, a shielding fabric for electrically conductive workwear, has a strength of 50 dB with a fibre composition of 80% spandex, 20% silver. The strength of Steel-Gray shielding fabric, designed for sewing overalls, is 35 dB, and the fibre composition is 40% cotton, 30% polyester, 30% stainless steel. ХНГ-80, ХНГ-100 - technical fabrics with metal fibre with universal protective coating, used in medicine and astronautics. Universal protective coating fabric HNG-100 has a reflection coefficient of 100 dB, contains 90% polyester, 10% copper. HNG-80 material has a reflection coefficient of 80 dB, fibre content of 80% polyester, 20% nickel. The protective coating strength of Silver-Silk metallic fibre technical fabric is 60 dB, it contains 80% nylon and 20% silver [39, 16-1966].

Today, metallic fibre fabrics are used in the production of special electrically conductive clothing, and the range of use is expanding year by year due to their properties. Much research is devoted to the use of electrically conductive fibres to create antistatic materials and reduce electrification of products [40-49].

There are various ways to reduce the electrical resistance of polymeric materials, one of which is to treat the surface of the fabric with an antistatic agent. The antistatic properties of the fabric created in this way are temporary and disappear when washing. Production of fabric with long-term antistatic properties is achieved by adding conductive fibres to its composition. In this case, electrically conductive fibres mixed with ordinary synthetic fibres in a certain ratio are used [42, 397-401 p.]. .

Metallic fibres can be used in the production of fabrics, knitwear and nonwovens. Metallic fibres are introduced into fabric by some textile methods, for example, clothing consisting of 25% metallic fibres from Bekinox® (Belgium) and 75% conventional textile fibres provides reliable protection against biological effects and side effects when working in electric fields [43, 1 p.]. In addition to the manufacture of special clothing, such fibres are also used in the production of antistatic carpets and special floor coverings.

However, at present metallised fibres are obtained by a very complex technology, have a high weight and a very high price. Therefore, today there is a tendency to replace metallised textile fibres, which have a number of advantages (lightness and cheapness). From the point of view of textile production it is convenient to work with metallised fibres as their physical and mechanical properties (strength, elasticity, stretchability) are very close to those of ordinary textile fibres [44. 48-49 pp.].

In some cases, only 0.5-1.0% of the total content of antistatic fibres is sufficient to give textile materials the required antistatic properties. The use of special clothes made of antistatic fabrics protects a person from static electricity [45, 184-190 p.; 46, p. 116-118. 47, 101-102 p.].

The American company Barnet provides antistatic properties of the fabric brand "NegaStat® " ***with*** electrically conductive thread, which is the most modern means of protecting people from the danger of electric shock when static electricity accumulates on the surface of clothes. The principle of operation is based on the inductive removal of static electricity from the surface of the garment and does not require the creation of permanent conductive circuits necessary for the effective action of steel and carbon antistatic threads. The favourable sides do not lose their antistatic properties under repeated washing [48, 1 -2 p.].

The Japanese firm "Kurar" produces protective suits with the use of metallised threads "Selmek", located in the fabric with a pitch of 2 mm, which allows to reduce many times the current passing through the human body in situations [49, 1 p.; 50, 47-50 p.].

Scientists of the Russian Federation have developed an electrically conductive heat-resistant fabric, which is made by weaving semi-conductive fibres with a surface porosity of 50-65%, a core yarn density of 16-20 threads/cm and a main yarn density of 10-20 threads/cm. Each warp and weft yarn is made of 350 - 400 twist/strand and combined yarns with a density of 90 - 120 yarns and complex polyaramid yarns with a density of 14.3 - 58 yarns. Twist - 100-200kr/m, mass fraction not more than 15%, diameter 70-120 microns, including copper nickel-plated wire. The ratio of copper and wire is in the range of 50-70:50-30 (RU №2054064 C1, D 03D 15/00, A41D 13/02, 10.02.1996). which has unsatisfactory sewing - consumer properties, limitations in the structure and design of the fabric, is not intended for sewing clothes and other complex products. This fabric can be used only for special coverings [51, 1-8 p.].

It is known that gauzes woven in the form of flowers contain a system of different warp and weft yarns, each of which includes chemical fibres forming the background and mesh yarns forming an electrically conductive mesh. In this case 8-10 kinds of threads are used, including 2 warp and weft threads each, every 2 or 3 threads of the background warp are connected with every 2 or 3 threads of the background warp, and the number of single weft threads is 1 or 2, from 4 to 8 rapport threads. (RU No. 2228398 C1, D 03 D 15/00, 10.05.2004) [52, p. 1-9]. Analogues of this type of fabric SU #1656020 A1[53, 1-7 lb.]; #2085635 C1[54. 1-8 p.], RU #2289642 S1 [55, 1-8 p.], RU2016929C1 [56, 1-11 p.], US #4606968A [57, 1-6 p.] are also cited in the patents.

The main disadvantage of the above fabrics is that the presence of metallised threads in their structure does not allow direct contact with the human body. In addition, the presence of metal in fabrics, both in the form of threads and in the form of coating, significantly increases the weight of the material, makes the production process more expensive and, as a consequence, increases the cost of antistatic fabric and product.

Another group of Russian scientists has created an electrically conductive antistatic fabric from complex electrically conductive polymer yarns of complex frame-core structure forming the basis of basic and jute yarns woven on the basis of chemical or natural fibres [55, 1-8 p.]. Complex electrically conductive polymer yarns of the shell-and-core structure of this fabric are made on the basis of capron yarns treated with a continuous electrically conductive composition on the basis of the polymer vinylidenefluorolite-trafluoroethylene filled with technical carbon in the ratio from 1:0,6 to 1:0,8. Thus the general linearity of electrically conducting polymeric fibres makes 52-58 tex (RU № 2289642 C1, D 03D 15/00, 20.12.2006). The disadvantages of this electrically conductive fabric include the use of carbon threads as electrically conductive threads, which due to the properties of such threads leads to various problems in the process of weaving on textile equipment (breakage, fragmentation, etc.) . In addition, the presence of softeners in the fabric sharply reduces the consumer properties of the fabric (sewing, wearing, washing) and thus limits its application area.

Polyester-cotton blend sari fabrics Premier Standard 180, Premier Standard 250 of the Russian company "Chaikovsky Textile" are designed for sewing special summer clothes and have high durability. These fabrics are resistant to oil and water, and in addition to strength and durability they have high hygienic properties due to the fact that they contain 80% cotton fibres.

Premier Cotton Rich and StretchAll 230 are also products of this company and are blended fabrics made of polyester-cotton fibre in sari-spun, designed for protection against general industrial pollution and mechanical effects. As the composition consists mainly of cotton fibre, it has high hygienic properties as well as oil and water resistance and is characterised by durability.

This company's Premier Cotton 330 fabric is made from 100% cotton fibre with a satin weave, which is not only highly hygienic but also dustproof and extremely low water permeability.

Thus, based on the above, we can conclude that antistatic fabrics are not produced in our country today. Since most antistatic fabrics produced abroad use metal threads, they are very expensive, and clothes made of these fabrics are heavy and uncomfortable to wear, which in this case is very important. From the point of view of superiority of hygienic, protective and operational properties and affordability of price, it was decided that the most suitable for workwear is antistatic fabric Premier Cotton 330, produced in the Russian Federation and widely imported into our country.

In our further researches we will define the question of carrying out a comparative analysis of physical-mechanical and hygienic properties of antistatic fabrics available on the textile market of Uzbekistan as one of the main objectives of the thesis work.

CHAPTER 7

FORMULATION OF REQUIREMENTS FOR ELECTRICALLY CONDUCTIVE PROTECTIVE CLOTHING

The analysis of the existing assortment of special clothing and the study of its protective ability as the main indicator of its quality is a complex and multifaceted task, and its relevance in modern conditions is beyond doubt.

Satisfaction of this set of requirements is provided, first of all, by the optimal combination of properties of materials from which the workwear is made and its structural and constructional elements. Thus, non-standard clothing should be developed mainly based on the requirements of consumers.

Overalls for electrical workers should be made according to the system of labour safety standards, but, unfortunately, currently most of the sewing enterprises of our republic produce overalls not according to the requirements of GOSTs, and focusing on the price policy. .

It allows us to analyse the working conditions of district power engineers, peculiarities of the thermal state of the human body in a hot climate and existing types of overalls, as well as to formulate requirements for manufactured overalls.

General Requirements:

- overalls must be a complete set, except for the jacket, made of antistatic fabrics with high hygienic properties;
- special clothing should sufficiently protect the human body from static electrical charges, sunlight and high air temperature [59];
- overalls should be fully adapted to working conditions, ensure physical freedom of movement of a person, and be comfortable to use;
- the design of special clothing should provide the necessary air exchange and at the same time limit the entry of electrical charges, thermal radiation and dust into the human body, as well as contribute to the timely release of appropriate moisture (sweat);
- overalls must be provided without service for 12 months under these conditions[59];
- special clothing should be modern, meet the requirements of the fashion trend, have a model composition and function, be clear and sufficiently expressive;

- overalls should be light: the total weight of the set should not exceed 1-2kg [59];
- overalls should be economical, i.e. made of cheap materials with minimal production costs, while ensuring their high quality.

Material Requirements:

- materials must not allow electrical charges and dust to enter the environment under the clothing [59];
- air permeability of materials should be at least 30 dm^3/m^2 s [61];
- The hygroscopicity of materials should be at least 7 % [62];
- The tensile strength of the materials should be more than 5-6kg [63];
- The wear resistance of the fabric should not exceed 2000 cycles, i.e. the wear period should not exceed 12 months [63];
- after washing, the warp threads of the fabric should not exceed 4 % and the warp threads should not exceed 2 % [64,65];
- fabric density does not exceed 200-300 g/m^2 [66].

Design Requirements:

- the design of special clothing must ensure normal human temperature;
- taking into account the topography of sweating, ventilation holes must be provided in the armpit area, along the basque lines;
- It is necessary to ensure that the design of overalls does not restrict the worker's movements and sufficiently covers his body when working;
- due to the physical and mechanical parameters of the fabrics, in accordance with the topography of bending and interaction of the elements, there must be protective coverings on the knees and lower parts of the legs;
- the design should be simple, the number of seams and parts should be brought to an average level, the cuts of the parts should be aligned as much as possible, and the parts themselves should be cut as a single unit, it is recommended to join the parts of metal parts located close to each other DRU^{Г} U,^{в} one complex assembly (combining pockets with coatings reinforcing the pockets, etc.);
- construction should be technological, i.e. ensure the use of modern equipment. It is

necessary to minimise moisture-thermal operations in the production process, use a simple assembly scheme and standard assemblies and parts for the manufacture of products.

Therefore, a programme of requirements for electrically conductive special clothing was developed taking into account the working conditions of district electric power engineers and the impact of hazardous and harmful production factors. Special clothing developed taking into account these requirements reliably protects workers from unfavourable climatic and harmful production effects.

CHAPTER 8

INITIAL STAGES IN THE DEVELOPMENT OF ELECTRICALLY CONDUCTIVE FABRICS

Jacquard knitted fabrics can be created on the basis of all main and derived weaves. Depending on this, the fabric can be:

- single;
- double;

Also:

- Culinary;
- basic knitting.

Types of jacquard weave

Coulier jacquard knits are usually produced on crochet machines with hooked needles. This type of fabric can also be produced with patterned culinary wheels. Often jacquard weaves are also produced on units with tongue needles. In this case, the method of including or not including the needles in the work is used.

Coulier jacquard knitted fabric can be single. It is in turn divided into two varieties:

- regular;
- irregular;

In a regular culinary jacquard weave, almost all the loops are the same size. In this case, each row of the fabric contains as many threads as there are colours for the future material.

For example, a two-colour knit has two threads, a three-colour knit has three threads, and so on.

Irregular knitting is characterised by large and small loops. Here, a row of loops is formed with one or more threads.

The single culinary jacquard knit is also available in shades.

It can be created by combinations of jacquard loops and ironwork. Relief single knits are also found in knitting. It is obtained by using jacquard loops with a high index (e.g. double, triple, etc.) as a basis. Jacquard loops compress the plain loops in such a way that bulges are formed on the surface of the fabric.

In addition to the single jacquard knit, there is a double jacquard knit. It comes in:

- complete;

- incomplete.

The full knit is made on a 1+1 eraser basis.

The needles for the front of the fabric work selectively, according to the pattern. At the same time, the needles used for the reverse side of the fabric function without exception.

When knitting an incomplete jacquard jacquard knit, the front needles perform the pattern and are partially included. At the same time, the circular needles perform the movements one after the other.

Often double jacquard knitwear is grouped according to the pattern. This classification is as follows:

- coloured irregular knitted fabric. Its face loops differ not only in colour but also in size;
- relief knit. Here, groups of jacquard loops are used on the front side, which have an index of 1, 2 or more. The jacquard loops pull the plain loops together. As a result, bulges are formed on the wrong side. They can have different shapes. For example, rhombuses, ellipses or plain strokes;
- An overlay knit of jacquard weaves. The fabric is formed by placing some threads on the front needles and other threads on the back needles. Notably, the front needles work exclusively on the pattern. The back side of this knit is capable of being monochrome, striped or having a staggered arrangement of loops. On the front side there are plain loops, which in certain areas are joined with jacquard loops;
- shaded jacquard weave knitwear. This jacquard weave of yarns is due to the combination of facing loops that are different in the height of the buttonhole row;
- jacquard weave fang knitted fabric. On the front side of the material there are jacquard loops, which are formed according to the pattern. On the reverse side there are plain loops in some areas and pressed loops with sketches in other areas.

It is worth saying a few words now about the basic knitted jacquard weaves.

To obtain it, the fibre is laid only on the needles from which the old loop is dropped. There are two types of weave: single and double.

The single warp knitwear of jacquard weaves is also endowed with its own classification. It is systematised into:

- embossed.
- downtime.
- embossed embroidery.

Double warp knit jacquard weaves are produced on tongue needles, when the fonts move frontally.

The fabric obtained from the theoretical and practical studies has the shape, in Figure 1.

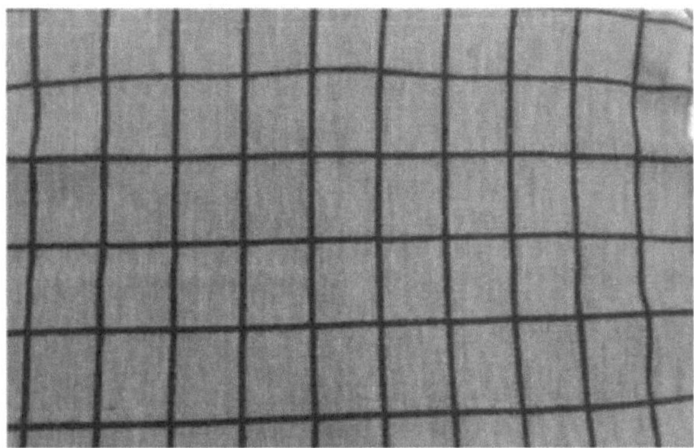

Figure 2. Developed prototype of electrically conductive fabric

In this fabric, the efficiency of electromagnetic shielding arises due to the fact that the fabric is a mesh of metal-containing (electrically conductive) threads, and metal mesh is known to be effective in shielding electromagnetic radiation (Electrodynamics of Mesh Structures. M.I.Kontorovich, M.I.Astrakhan, V.P.Akimov, etc. / Edited by M.I.Kontorovich. - M.: Radio and Communication, 1987).

The technical problem, which the claimed invention is aimed at solving, is to expand the assortment possibilities by producing fabrics that reduce the intensity of external electromagnetic radiation while preserving sewing and consumer properties

The technical result is the improvement of sewing and consumer properties of conductive shielding fabric.

Earlier developed conductive fabrics for sewing covers for apparatuses and devices were very inconvenient for quality sewing, because they were not intended for sewing clothes, so we had another task to create a quality fabric designed specifically for sewing clothes, so several types of fabrics were developed.

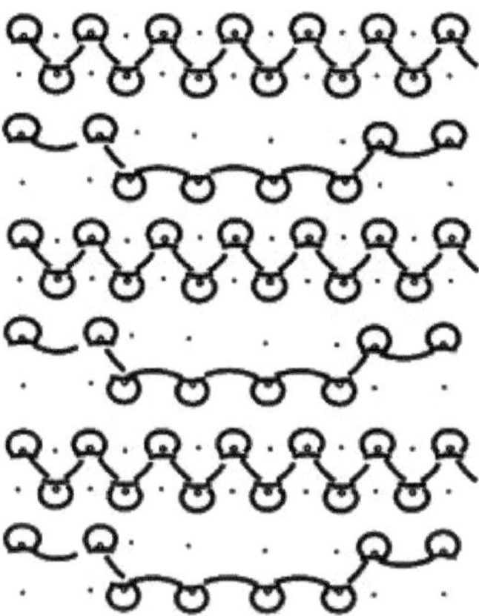

Figure 3. Schematic of the fabric

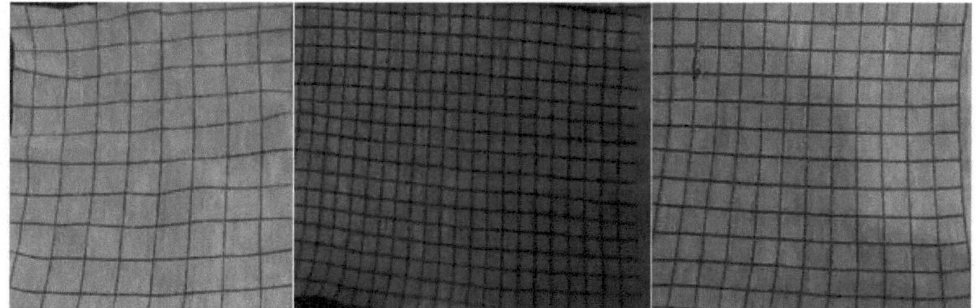

Figure 4. Prototypes of the developed material with difference in patterning

These prototypes differ from each other in that the pattern of this fabric is created from the interweaving of ordinary yarns and yarns with conductive metal. That is, the light part of the fabric consists of ordinary knitted yarns, and the dark threads are those electrically conductive threads with metal in their composition.

It was assumed that a fabric with a large electrical mesh gap has a smaller conductive function and one with a fine gap, i.e. fine mesh pattern has a larger conductive function.

To validate the hypothesis, laboratory tests of the fabric for electrical conductivity were carried out and the results are recorded in Table 1:

Table 1:

Volt Ampere characteristic of an electrically conductive filament

Вольт Ампер характеристика електропроводящей нити																	
№	V	4 ряд			6 ряд			8 ряд			10 ряд			12 ряд		
		I	II	III	I	II	III	I	II	III	I	II	III	I	II	III
1	0,5	0,4	0,8	0,3	0,4	0,2	0,1	0,7	4,2	5,1	7,2	2,9	1,8	1,5	3	5,2
2	1	0,7	1,8	0,7	0,7	0,7	1,1	2,2	8,1	10,2	14,3	5,1	4,3	3,5	5,4	10,6
3	1,5	1,1	3,2	1,7	1,3	3,3	2,4	4,2	12,6	14,2	19,2	7,3	7,8	5,6	8,3	10,9
4	2	1,5	4,6	2,7	3,1	4,6	4	7,1	17,8	18,8	25,2	9,7	12,3	8,8	11,7	20,3
5	2,5	2,1	6,3	4,2	4,7	6,6	8,5	9,2				11,8	16,8	13,7	155,3	24
6	3	2,8	8,5	5,5	10,1	8,9						12,8	25		20	33,8
7	3,5	3,8	8,5	60		10							76			
8	4	5,4	11,3			13										
9	4,5	7,5	90,6			100										
10	5	9,1														
11																

Thus, to date, the jan topic has been fully explored in all relevant respects and most of the work has been done.

As the topic is very actual for today, it is planned to solve this problem as soon as possible and to choose the most optimal prototype from the created prototypes based on the results of laboratory studies of physical and mechanical properties of the created materials.

Their physical and mechanical properties are the most important for isolation of accurate results of the whole research and creation of a trial model of special clothing for further testing of the properties of clothing, in conditions maximally equated to the present working conditions with electromagnetic field emitting electromagnetic waves.

List of references used

1. GOST 12.1.002-84 System of labour safety standards. Electric fields of industrial frequency. Permissible levels of intensity and requirements for control at workplaces. - Moscow: Standardinform, 2009. - 7 c.

2. Pulatova S.Y., Kadirova S.H. Analysis of nomenclature of materials intended for overalls against electromagnetic radiation. Development of science and technology. Scientific and technical journal. Bukhara, 2019. № 4 C. 118-123. (05.00.00; №24).

3. RU2006108024/12A 2006-03-14 Conductive fabric

4. https://www.aramilgo.ru/news/rospotrebnadzor/6137-vliyanie- elektromagnitnyh-izlucheniy-na-zdorove-cheloveka

5. Decree of the Cabinet of Ministers of the Republic of Uzbekistan from 04.10.2017 № 795. On further improvement of the classification of the main positions of employees and occupations of workers.

6. Abel J. Gonzalez, Makoto Akashi, John D. Boyce Jr, Masamichi Chino, Toshimitsu Homma, Nobuhito Ishigure , Michiaki Kai, Shizuyo Kusumi, Jai-Ki Lee, Hans-Georg Menzel, Radiological protection issues arising during and after the Fukushima nuclear reactor accident https ://doi.org/10.1088/0952-4746/33/3/497 . pp. 497-571.

7. Castillo J., Cubillos A. Ergonomic principles of workwear design for electricians in Colombia/WORK-A Journal of Prevention Assessment & Rehabilitation. - 2012.- Issue. 41.- C. 623627.

8. Lin LX, Wang YY, Li J. Functional design and ergonomic evaluation of earthquake search and rescue clothing/7th Textile Bioengineering and Informatics Symposium (TBIS 2014)/5th Asian Protective Clothing Conference (APCC 2014) Proceedings: - 06 Aug - 08, 2014. - C. 348-355.

9. Azhar Ali, Patrick Schamberger and Sriram Vaddiraju. Thermal conductivity of ensembles of Mg2Si1-xSnx nanowires synthesised by solid-phase transformation of silicon nanowires. Materials Research Express , volume 8 , number 2 . pp . 2814-2819.

10. Romanov V.E. System approach to design of special equipment. - Moscow: Light Industry, 1981. - 128 c.

11. Surzhenko E.Ya. Theoretical bases and methodical support of ergonomic design of special clothes: autoref . . Doctor _ tech. date: 19.05.04. - SPb: SPGUTD, 2001. - 49 c.
12. Silchenko E.V. Development of fabrics for special clothes against electromagnetic radiation : a disc . sugar . tehn. sciences: 19.05.02 . - Moscow : Kosygin Russian State University - 20 1 8.- 147 p.
13. CherunovaI . V. Theoretical bases of complex designing of special heat protective clothing: dne .... Doctor of Tech. date: 19.05.04. - Shakhty: Yurges - 2008. - 307 c.
14. Merkulova AV. teeth _ . candy. tech. date: 19.05.04. - Shakhty: Yurges - 2007. - 184 c.
15. Kurenova I.V. Development of research of special oil protective clothing with a modified package of materials: autoref . . conf. tech. date: 19.05.04. - Shakhty: Yurges, - 2013. - 28 c.
16. Ivashchenko I.N. Development of special clothes for protection of workers of oil-producing industry of the Southern region of Russia from low temperatures: dne . . conf. tech. date: 19.05.04. - Moscow: Tsentr. Research Institute of the garment industry, 2008.- 199 p.: ill. RGBOD, 61 08-5/1556
17. Shultz L.O. Perfection of methodical and instrumental support of the process of designing overalls against increased temperatures: autoref . dis . can. technology science - Leningrad. - 1989. - C. 3-8.
18. Akbarov D.N., Baimuratov B.H., Akbarov R.D. Giving antistatic properties to textile materials //" in Uzbekistan light industry innovations based on the development of the current issues " Materials of the national conference . - T., 2012. C. 97-99.
19. Akbarov R.D. Development and research of properties of special fabric with given electrophysical characteristics: author's thesis. can. technology science - Tashkent, 2005,19 p.
20. BaimuratovB .H. Creation of multifunctional conductive textiles materials and their properties 05.0 6.02 - Submission for awarding the degree of

Doctor of Sciences on the basis of a patent for invention without defending a dissertation in the speciality of textile materials technology and raw material preparation .

Tashkent - 2020.

21. Gusev B.N., Matrokhin A.Yu. Textile material science before the technological breakthrough // Izvestiya vysshee obrazovaniya vysshee obrazovaniya. Technology of textile industry. - 2016. - № 1 (361). - C. 42-47.

MIX
Papier aus verantwortungsvollen Quellen
Paper from responsible sources
FSC® C105338

Printed by Books on Demand GmbH, Norderstedt / Germany